ADDISON-WESLEY

QUEST 2000

EXPLORING MATHEMATICS

AUTHORS

Ricki Wortzman · Lalie Harcourt · Brendan Kelly · Peggy Morrow

Randall I. Charles · David C. Brummett · Carne S. Barnett

CONTRIBUTING AUTHORS

Linda Beatty · Anne Boyd · Fred Crouse · Susan Gordon

Elisabeth Javor · Alma Ramirez · Freddie Lee Renfro · Mary M. Soniat-Thompson

REVISED EDITION

Addison-Wesley Publishers Limited

Don Mills, Ontario · Reading, Massachusetts · Menlo Park, California
New York · Wokingham, England · Amsterdam · Bonn
Sydney · Singapore · Tokyo · Madrid · San Juan · Paris
Seoul · Milan · Mexico City · Taipei

REVIEWERS/CONSULTANTS

Marie Beckberger, Springfield Public School, Mississauga, Ontario
Jan Carruthers, Somerset and District Elementary School, King's County, Nova Scotia
Garry Garbolinsky, Tanner's Crossing School, Minnedosa, Manitoba
Darlene Hayes, King Edward Community School, Winnipeg, Manitoba
Barbara Hunt, Bayview Hill Elementary School, Richmond Hill, Ontario
Rita Janes, Roman Catholic School Board, St. John's, Newfoundland
Karen McClelland, Oak Ridges Public School, Richmond Hill, Ontario
Betty Morris, Edmonton Catholic School District #7, Edmonton, Alberta
Jeanette Mumford, Early Childhood Multicultural Services, Vancouver, B.C.
Evelyn Sawicki, Calgary Roman Catholic Separate School District #1, Calgary, Alberta
Darlene Shandola, Thomas Kidd Elementary School, Richmond, B.C.
Elizabeth Sloane, Dewson Public School, Toronto, Ontario
Denise White, Morrish Public School, Scarborough, Ontario
Elizabeth Wylie, Clark Boulevard Public School, Brampton, Ontario

TECHNOLOGY ADVISORS

Fred Crouse, Centreville, Nova Scotia; Flick Douglas, North York, Ontario; Cynthia Dunham, Framingham, MA;
Susan Seidman, Toronto, Ontario; Evelyn J. Woldman, Framingham, MA; Diana Nunnaley, Maynard, MA

Editorial Coordination: McClanahan & Company
Editorial Development: Susan Petersiel Berg, Margaret Cameron, Mei Lin Cheung,
Fran Cohen/First Folio Resource Group, Inc., Lynne Gulliver, Louise MacKenzie, Helen Nolan, Mary Reeve

Design: McClanahan & Company
            Wycliffe Smith Design

Cover Design: The Pushpin Group

Canadian Cataloguing in Publication Data

Wortzman, Ricki
        Quest 2000 : exploring mathematics, grade 2
revised edition : student book

First and third authors in reverse order on
previous ed. ISBN 0-201-55263-9

I. Mathematics – Juvenile literature.   I. Harcourt,
Lalie, 1951–   . II. Kelly, B. (Brendan),
1943–   . III. Title

QA107.K43 1996   510   C95-932764-9

ISBN 0-201-55263-9

This book contains recycled product and is acid free.

Printed and bound in Canada.

    D E F — FP — 01 00 99 98

Table of Contents

How can we use data to tell about us?

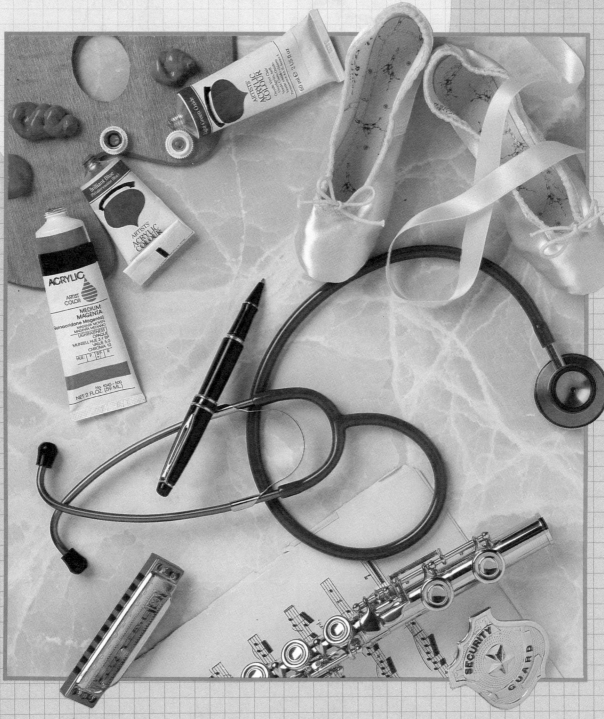

🏠 **FAMILY NOTE:** For about the next four weeks, your child will be exploring ways to ask questions of others and ways to present the information he or she finds. This is part of the children's work with collecting and organizing data. You may wish to help your child think about what questions he or she would like to ask classmates.

Name _____

Just Like Us

How are these children alike?

What would you draw to help this child
fit in the group?

🏠 **FAMILY NOTE:** The children have been sorting themselves into groups based on
likenesses. Ask your child to share the rule he or she used for sorting the last child
into the group.

I Can Tell

Some people use these things.
What can you tell about people who use them?

Draw a picture that tells others something
about you.

Ideas for Sorting

Here are some ways to label groups.

Ride a Small Bike	Ride a Big Bike	Play Soccer	Play Baseball
Like to Roller Skate	Like to Ice Skate	On a Soccer Team	On a Baseball Team
Like to Sing	Like to Play an Instrument	Play Kickball	Jump Rope
Play Games	Play Football	Like to Draw	Like to Read

Name _____

Where do you belong?

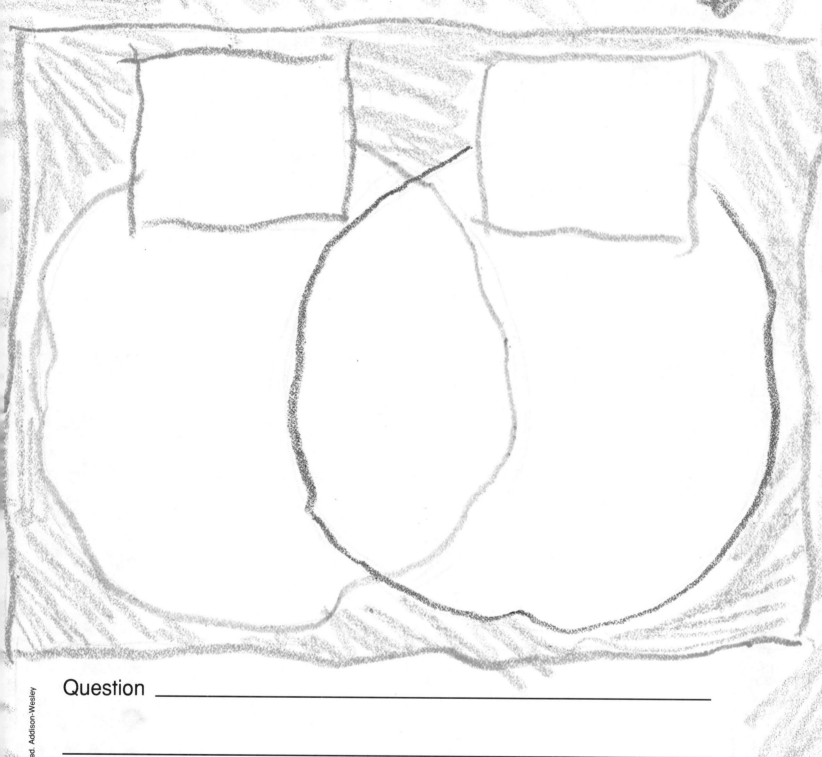

Question _____

_____

FAMILY NOTE: The children have been learning that they may be a member of more than one group. Ask your child why a specific name is written where it is. Help your child to add your name.

SHARING SIMILARITIES

5

Name _____

Name Game

These are ways I sorted my friends' names into groups.

🏠 **FAMILY NOTE:** Your child has been finding ways to put some names together to form a group. Ask your child to tell how the names in each group are the same. Help your child add names of family members to any of the groups.

What's My Shape?

Do you think you are shaped like a square?
Do you think you are shaped like a rectangle?

Measure to find out.

Use string to find
your height.

Use more string to
find your arm span.

Compare the lengths of string.

What shape are you? Tell how you found out. _____

_____

_____

🏠 **FAMILY NOTE:** Your child measured her or his height and arm span to see if her/his shape forms a square or a rectangle. Ask your child to explain how she or he decided. Your child may enjoy measuring the rest of the people in your family.

Name _____

Questions, Questions

Here are some questions you might choose to use for your survey.

1. What colour are your eyes?

brown   blue   hazel

2. Which do you like best?

  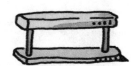

swings   slide   climbing bars   balance beam

3. Which of these ice-cream flavours do you like best?

vanilla   chocolate   strawberry

4. Which sport do you like best?

baseball   basketball   kickball   soccer

5. How many people are in your family?

2   3   4   5   more than 5

🏠 **FAMILY NOTE:** Your child has been working to ask survey questions of classmates. Help your child complete these survey questions and then ask the questions of family members.

Name _____

My Survey

Question _____

_____

What I learned from taking this survey: _____

_____

_____

Name _____

Surveys with String

What ways do you record what you count?
What special tools do you use to help you?

The Inca people lived in South America.
They used strings like these to record what
they counted.

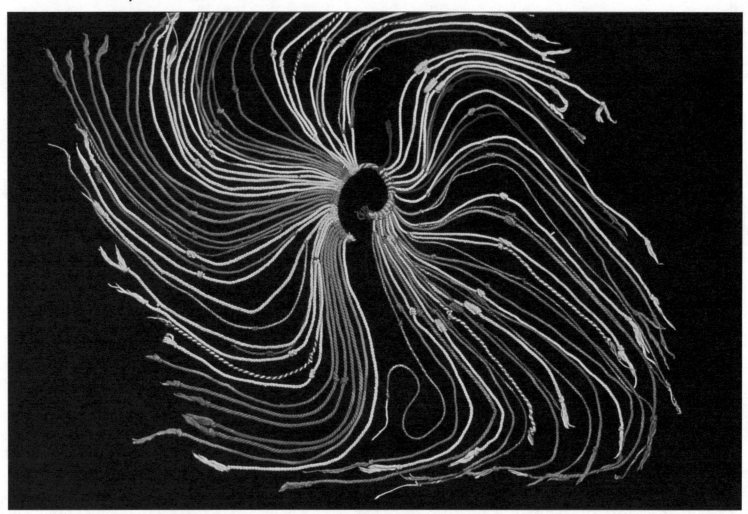

They made knots on strings to keep count of
things. They counted how many children were
born and how many animals they had.

How would you make knots to show how many
people are in your family?

Our Favourite Things

Favourite Recess Game

Tag ||||
Ball ||||
Races ||||
         |

Favourite Recess Game

Tag ☺☺☺☺☺☺☺☺☺☺☺

Ball ☺☺☺☺☺☺☺☺☺☺☺☺☺☺

Races ☺☺☺☺☺☺☺☺☺☺☺

Favourite   Recess   Game

Tag
Paul
Steven
Lori
Amy
kyle
Derek
Susan
Malik
Juan
Tonya
Ching
Jann

Ball
Gabriel
Janice
Luis
Amber
David
Karen
Maria
Tran
Sara
Kevin
Katie

Mario
Chandra
Len

Races
Teresa
Micheal
Jay
Rebecca
Billy
Liz

Name _____

Our Favourite Things

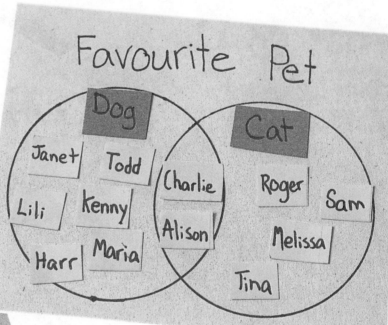

Think About THIS

Imagine trying to count every person in the country!

This person is helping to do just that! What question might she ask?

How do we solve problems with numbers?

🏠 **FAMILY NOTE:** Your child will be spending about four weeks using addition and subtraction to solve problems. We will focus on looking for and identifying patterns in facts to help us remember them. Ask your child to tell how he or she sees addition and subtraction being used in these pictures.

Name _____

Telling Number Stories

🏠 **FAMILY NOTE:** Your child has been working to make up adding and subtracting stories. Have your child share with you some of her or his stories and the addition and subtraction sentences that tell about the stories.

Line Up!

What numbers can you find?

What stories did you tell?

Taking a Ride!

What different ways could 10 children sit?

Write a number sentence to show each way.

FAMILY NOTE: We have been working to find all the possible ways that 10 children can be seated on the ride. We have been writing number sentences to show the ways. You may wish to have your child show you her or his ways using any counters that you have available.

Have a Seat!

What different ways could 15 children sit?

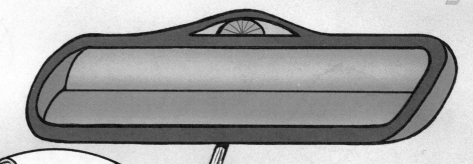
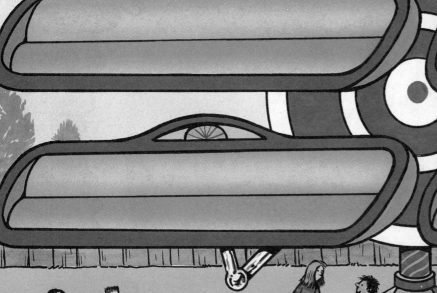

Think About THIS

For a long time people have been riding wheels like this one for fun.

This wheel was turned by hand. How many cars do you see on this wheel? How many people might fit in each car?

NUMBER COMBINATIONS

Name _____

Target Sums

2 4 6 8

Write a number sentence to show each total.

🏠 **FAMILY NOTE:** We played a game by dropping two counters onto the target, then adding the numbers to find the total. We also solved the problem of finding all the possible combinations and totals using two counters. Have your child show you how to play, then play a few rounds together.

Name _____

Make Your Own Target Board

Write a number sentence to show each total.

Name _____

More Target Sums

5 6 7

Write a number sentence to show each total.

🏠 **FAMILY NOTE:** We played a game by dropping three counters onto a target board. Then we added the numbers on which the counters landed to find the total. We also solved the problem of finding all the possible totals using three counters. You may wish to have your child show you how to play and play a few rounds together.

Name _____

Make Another Target Board

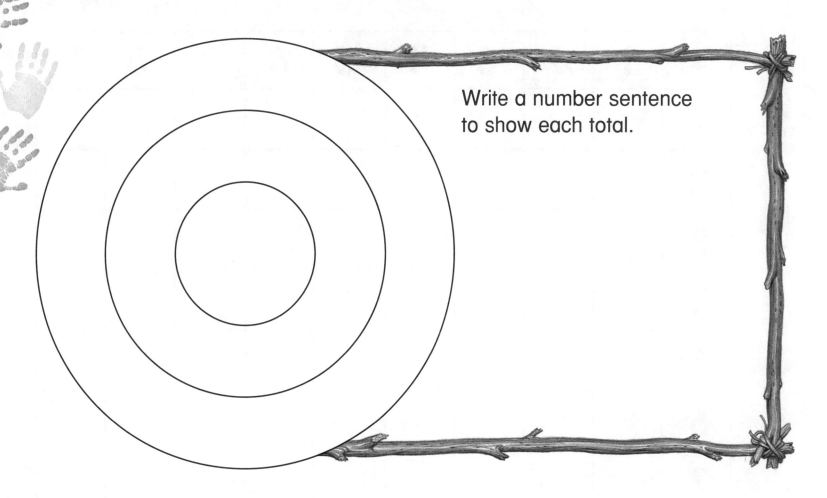

Write a number sentence to show each total.

Some people play a target game called bowling. There are 10 pins. You get 2 chances to knock them down.

What are some ways you could knock down all 10 pins?

First ball	Second ball

How many other ways are there? _____

Name _____

Addition Table

+	1	2	3	4	5	6	7	8	9
1									
2									
3									
4									
5									
6									
7									
8									
9									

Write about the patterns you see.

FAMILY NOTE: Your child has been working to find all the addition facts and sums for 2 through 18. Ask your child to describe how he or she found the sums and completed the chart. Ask your child to describe any patterns he or she found in the chart.

Math Tools

Here are two machines people use to add and subtract. One is very old and is called an abacus. One is newer and is called a calculator.

— Stands for 5

— Bar

— Stands for 1

Get a calculator.
Press [C] to start.

Enter an addition rule, like this [+] [6] [=].

Then press any number you want, and [=].

What happens?

Try another number and [=].

— Subtract

— Add

— Equals

Record your tries.   **My Rule** [+] [6] [=]

Number	Result

Let someone else press a number and [=].
Can they guess your rule?

FUNCTION
MACHINES

🏠 **FAMILY NOTE:** Your child has been using addition and subtraction to find relationships in sets of numbers. For example, for the numbers 2 and 5, 3 and 6, you add 3 each time. This page shows an example of a tool used to add.

Smart Shopper

Write about where you would buy each item.
Tell how much money you would save.

Where should people shop? Why?

FAMILY NOTE: Your child is using addition and subtraction to solve problems involving comparing prices and finding how much money can be saved on each item. Have your child tell you how he or she solved this problem.

Target Totals

How to Play

1 Take turns.

2 Drop 3 counters.

3 Find your total.

4 The higher total wins.

Find how many
the winner won by.

The winner scores
that many points.

Play 5 rounds.

Add the points.
Who won and
by how many points?

4 6 8

Points

FAMILY NOTE: We played a game using three counters and dropping them onto the target board. Then we added the three numbers to find the total. Then we made up our own games. You may wish to have your child show you how to play and play a few rounds.

Name _____

Design Your Own Target Game

Things to Think About

- How many turns do the players get?
- How do players get their markers onto the target?
- How do you score points for each round?
- How many rounds do you play?

My Rules and How to Play

How do we spend our time?

🏠 **FAMILY NOTE:** For the next week, your child will be learning about time and the calendar. Ask your child to tell you about this picture.

27

UNIT
3
ACTIVITY
1

Special Events for My Calendar

Name _____

Choose some special events to show
on your calendar.
Here are some events you might
want to choose.

family vacation

puppet show

visiting Grandmom

Canada Day Holiday

EVENTS
IN
A
YEAR

28

🏠 **FAMILY NOTE:** Your child has been learning how to write events on a calendar.
Use the large sheet your child has brought home. Help your child write the dates of
any special days your family observes.

Earth Day

losing a tooth

new baby in family

school bake sale

List some of the events you would
like to show on your calendar.

_____   _____

_____   _____

_____   _____

_____   _____

Name _____

It All Takes Time

What can you do in 1 minute?
What can you do in 5 minutes?

🏠 FAMILY NOTE: Your child has been learning how long 1 minute is and how long 5 minutes is. Ask your child to tell you which of the activities above could be done in 1 minute and which would take 5 minutes or longer. Choose one activity and time your child as he or she does it.

How Do You Spend Your Time?

How much time do you spend in school?

I arrive in school at: I leave school at:

I spend _____ in school.

How much time do you spend sleeping at night?

I go to sleep at: I wake up at:

I spend _____ sleeping at night.

Write something about what you found.

🏠 FAMILY NOTE: Your child has been investigating how he or she spends time. Help your child keep a record of the time he or she goes to sleep and wakes up. Observe as your child draws hands on the clock to show each time.

Facts About Time

60 seconds are in 1 minute.

60 minutes are in 1 hour.

24 hours are in 1 day.

365 days are in 1 year.

Write about or draw an interesting
way you spend your time.

Think About THIS

A girl played hopscotch for
24 hours without stopping.
She finished 307 games.
That's one game
every 4 $\frac{1}{2}$ minutes!

What kinds of shapes can we make?

UNIT 4

Recognizing and Making Shapes

FAMILY NOTE: For about the next three weeks, your child will be learning about geometric shapes and solids in his or her world. Ask your child to tell you about the shapes he or she sees on this page.

Shapes and Solids

These are some shapes.

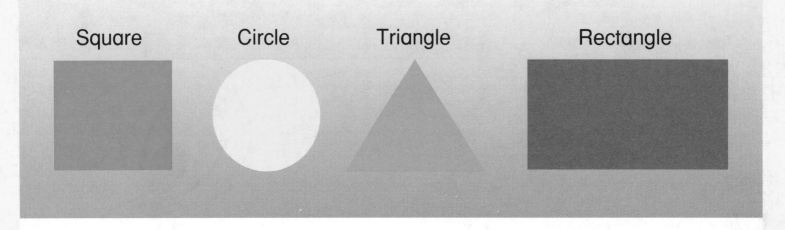

| Square | Circle | Triangle | Rectangle |

These are some solids.

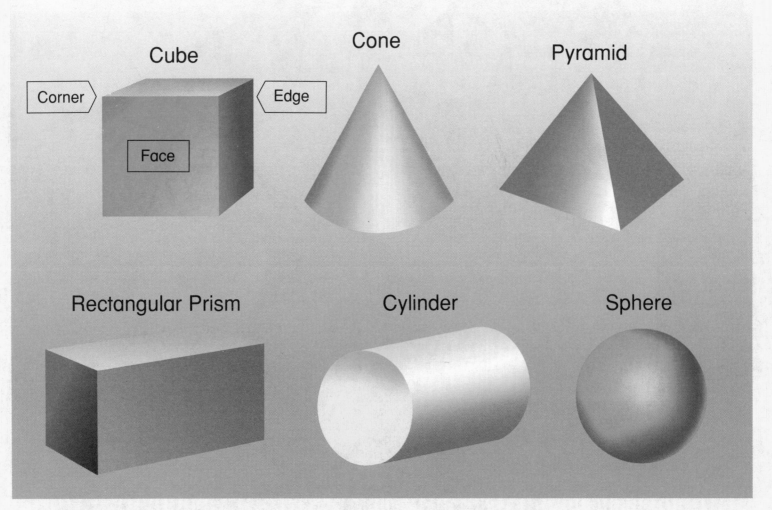

Cube

Corner Edge

Face

Cone

Pyramid

Rectangular Prism   Cylinder   Sphere

🏠 **FAMILY NOTE:** Your child has been investigating geometric solids and shapes. Look around at home with your child to see where these solids and shapes are found.

Opening Up Boxes

This is what I think my box will look like when I cut it open.

I cut open my box.
This is what it looked like.

These are the shapes I saw when I opened my box.

Think About THIS

Junkanoo is a festival in the Bahama Islands. For this festival, people decorate costumes to wear in a parade. This costume was made from an empty box. What can you do with an empty box?

FAMILY NOTE: Your child is learning about geometric solids. Your child examined and cut open a box to discover the shapes of all the sides. Encourage your child to tell you about the activity, and what he or she found out.

Name _____

The Tangram

This is a tangram.

A tangram is an ancient Chinese puzzle.

A tangram square is divided into 7 shapes.

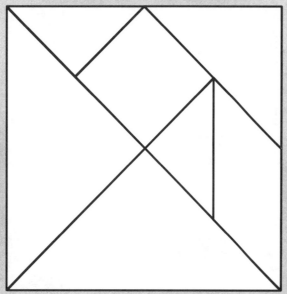

Write about or draw the shapes you see.

🏠 **FAMILY NOTE:** Your child has been investigating shapes that form other shapes. We have used a tangram puzzle in our investigation. To make a tangram at home, cut a 8-cm square from paper. Fold and cut the square into seven shapes as described on page 37. Ask your child to show you what he or she can make with the tangram pieces.

Making a Tangram

Make your own tangram.
You need a 8-cm paper square.

---

1. Fold the square.
Cut.

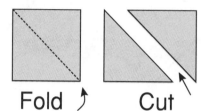

Fold ↗   Cut

---

2. Take one triangle.
Fold. Cut.
These are 2 puzzle pieces.

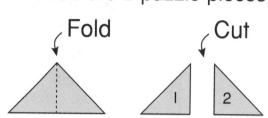

---

3. Take the other triangle.
Fold. Open.
Fold down the top.
Cut. This is puzzle piece 3.

---

4. Cut.

Cut

---

5. Take one piece.
Fold it to form a square.
Cut to form a square and a
triangle.
These are puzzle pieces 4 and 5.

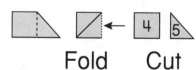

Fold   Cut

---

6. Take ⬭.
Fold. Cut.
These are puzzle pieces 6 and 7.

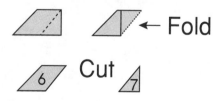

Name _____

Tangram Puzzles

Make each shape with your
tangram pieces.
Trace around each shape you make
on another paper.

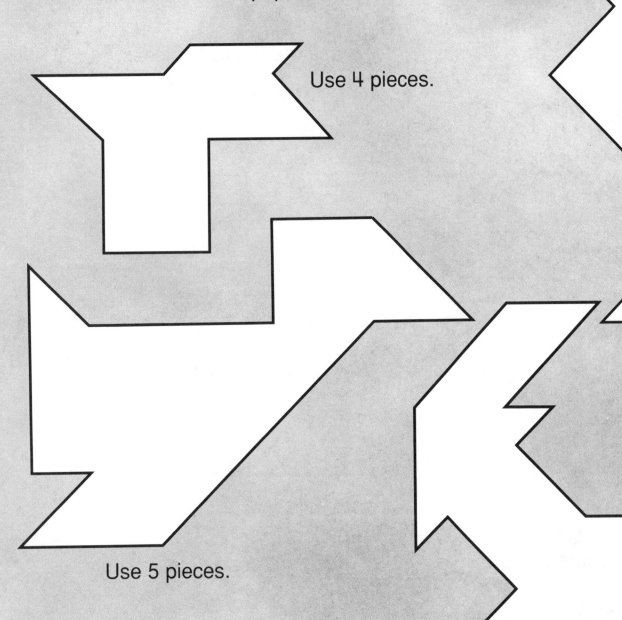

Use 4 pieces.

Use 6 pieces.

Use 5 pieces.

Use 7 pieces.

Growing Square Patterns

This square pattern keeps growing.

Make it with  Pattern Blocks.

1   2   3

What would the next square in the pattern look like?
Try it.
Write about your pattern. Use numbers to tell about it.

Number	Number of blocks in bottom row	Number of blocks in square
1		
2		
3		
4		

How many [block] would you use in the next square?
Guess. Then build it to check.

FAMILY NOTE: Your child is learning about patterns that grow. The square pattern above is a growth pattern. Ask your child to tell you about growing patterns, and what would come next.

Growing Triangle Patterns

This triangle pattern keeps growing.

Make it with Pattern Blocks.

1   2   3

What would the next triangle in the pattern look like?
Try it.
Write about your pattern. Use numbers to tell about it.

Number	Number of blocks in bottom row	Number of blocks in triangle
1		
2		
3		
4		

How many would you use in the next triangle?
Guess. Then build it to check.

House Shapes

On the plains, many of the First Nations people lived in teepees.

Which shapes do you see?

Some people in Africa live in houses like this one.

What shapes do you see?

🏠 **FAMILY NOTE:** Your child has been working to design and build a home using given shapes, sizes, and patterns. Ask your child to tell you about the home that he or she designed, and how he or she used shapes, sizes, and patterns in the design.

Name _____

More Shapes

Egyptians of long ago made
buildings like this one.

Which shapes do you see?

Today many people
work in buildings
like these in big cities.

Which shapes do you see?

How do we make big numbers?

🏠 **FAMILY NOTE:** For about the next four weeks, your child will be exploring numbers to 1000. The children will investigate larger numbers, and have opportunities for counting and writing. Ask your child to tell you about how many seeds he or she thinks are in this picture, and how the seeds could be counted.

Name _____

Collecting 1000

Dear _____

🏠 **FAMILY NOTE:** Your child has been exploring 1000. Help your child think of things that you might have 1000 of at home. Then ask your child to tell you about our class project.

How 1000 Looked Long Ago

Long ago in Egypt, people recorded
1000 like this:

1000 loaves of bread

1000 cattle

1000 geese

1000 gold coins

How do you think the Egyptian people
would show 2000? 3000?

Suppose you had to invent a way to show 1000.
What would you do?

Name _____

Inventory Control Sheet

Container	Our Estimate	Our Count	How We Counted

🏠 **FAMILY NOTE:** Your child has been counting collections of objects. Ask your child to tell you how many objects she or he counted and show you how she or he counted them.

Name _____

It's in the Bag

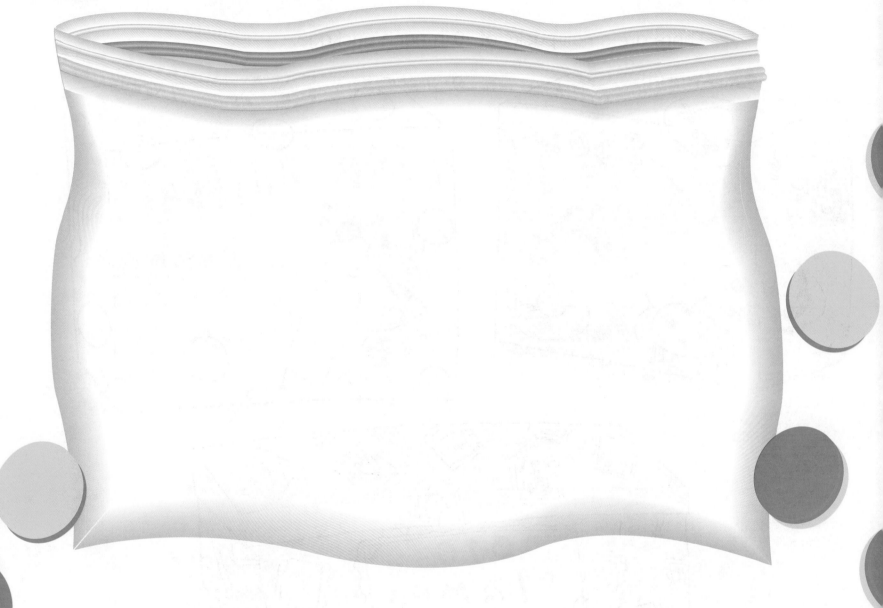

In my bag, you can see there are _____.

You can tell because _____

_____

_____.

🏠 **FAMILY NOTE:** Your child chose a way to group a large collection of objects to make them easier to count. Make a collection of 100 to 200 small objects at home. (Pennies or macaroni or dried peas work well.) Let your child group the collection and count the objects. Then, you group the collection and count. Did you and your child group the objects in the same way? Talk about it.

Name _____

Packs of 10

There are small packs of 10 in each package.

Choose a package.

How many small packs of 10 are there? _____

Show how you can prove it. Use another piece of paper.

⌂ FAMILY NOTE: Your child has been investigating how many tens there are in 100 and in other large numbers. Have your child choose one of the packages above. Ask your child to tell you how many tens there are in that number. Have your child show how he or she knows that.

Play a Target Game

Drop a counter on the target 5 times.
Record your scores on Activity Master 30.
Write your scores in order.

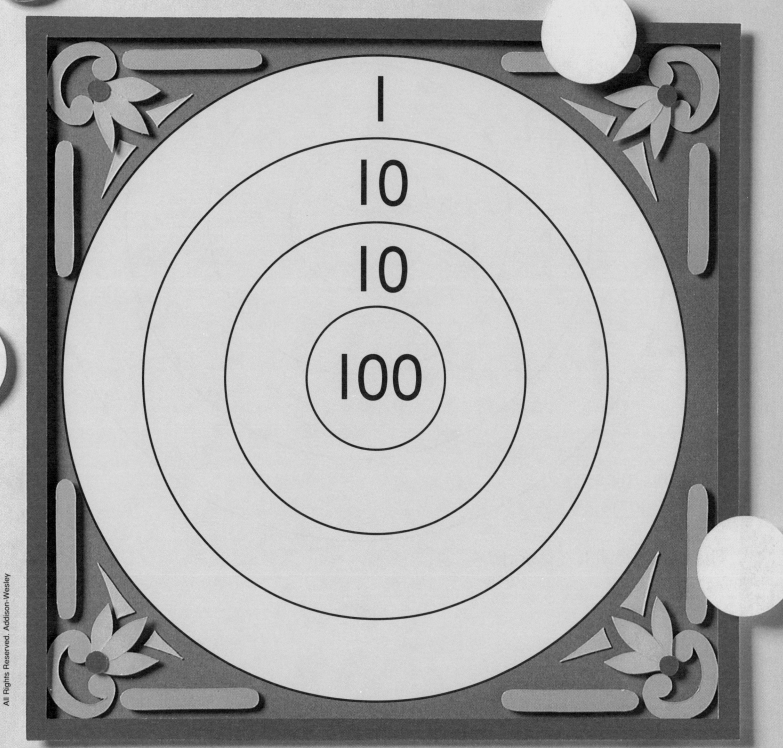

🏠 **FAMILY NOTE:** Your child played this target game with a counter. To play, each player drops the counter on the target five times, then finds the total score. Compare the scores. Arrange the scores in order.

Make a Target Game

Create your own target game.

Write the rules for your game.

Then play it with a friend.

Name _____

Counting Patterns

Colour a counting pattern.

1	2	3	4	5	6	7	8	9	10
11	12	13	14	15	16	17	18	19	20
21	22	23	24	25	26	27	28	29	30
31	32	33	34	35	36	37	38	39	40
41	42	43	44	45	46	47	48	49	50
51	52	53	54	55	56	57	58	59	60
61	62	63	64	65	66	67	68	69	70
71	72	73	74	75	76	77	78	79	80
81	82	83	84	85	86	87	88	89	90
91	92	93	94	95	96	97	98	99	100

My pattern counts by _____.

These numbers continue the pattern _____

_____

_____

_____

 FAMILY NOTE: Your child has been investigating skip-counting patterns to 100. Ask your child to share the counting patterns he or she coloured on the hundred chart.

EXTENDING
COUNTING
PATTERNS

Name _____

1000 Ideas

These are some ideas to help you with your poster on 1000.

Indian Brahmi

Hebrew

ONE THOUSAND

How many tens in one thousand?

How many hundreds in 1000?

How many ways can you make 1000?

How many people would make 1000 legs?

How many people would make 1000 fingers?

How many people would make 1000 noses?

How many dimes make 1000 cents?

How many nickels make 1000 cents?

Japanese

🏠 **FAMILY NOTE:** Your child has been working to design and make a poster to show what he or she knows about the number 1000. Have your child tell you about the things that he or she placed in the poster.

How do we add and subtract larger numbers?

🏠 **FAMILY NOTE:** For about the next four weeks, your child will be continuing to explore addition and subtraction as we develop procedures for adding and subtracting larger numbers. Ask your child to tell you about the pictures and talk about when adding or subtracting might be used in these scenes.

Name _____

Pencil and Eraser Sets

20¢

39¢

45¢

30¢

25¢

What sets can you make?
How much does each set cost?

Which set would you choose to buy? Why?

FAMILY NOTE: We have been working to find all the possible ways to make a set of one pencil and one eraser, using the items pictured on this page. Your child has been working to find the total cost of each set. Have your child tell you the way he or she found.

Piggy Bank Totals

20 pennies

9 pennies

37 pennies

15 pennies

What pairs of banks can you make?
How much money is in each pair of banks?

FAMILY NOTE: We have been working to find all the possible pairs using the banks
pictured on this page. Your child has been working to find the total amount of each
pair. Have your child tell you about the pairs he or she found.

Penny Savings

Suppose you and a friend are saving pennies.
How many pennies have you saved?
How many pennies has your friend saved?

_____ pennies

_____ pennies

How much do the two of you have in all?

Multicultural Games

Games with cards, dominoes, and dice
are played all over the world.

Playing Cards from India

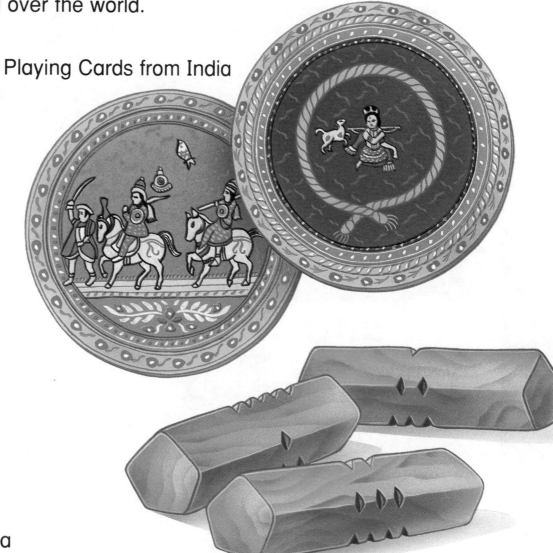

Domino Cards from China

Five-Sided Dice from Korea

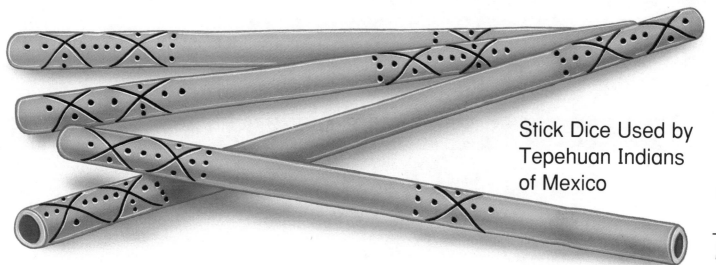

Stick Dice Used by
Tepehuan Indians
of Mexico

Name _____

Design a Number Game

You can make your own game.

Think about:
- What number do you plan to hit?
- Do you have to do it in one turn?
- Is there a rule for choosing the starting number?

My Rules and How to Play

 FAMILY NOTE: Your child has been playing a game called Getting Close to 100 to help her or him begin thinking about adding two large numbers such as 36 and 45. Have your child tell you about the game he or she designed.

Shopping Bargains

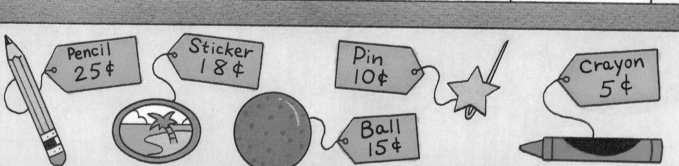

What will you buy?
How much money do you have left
after you buy each item?

🏠 **FAMILY NOTE:** Your child has been working to find how much money is left after purchasing items that he or she selects. Have your child tell about the method he or she used to find out.

Name _____

Your Store

_____'s

Bargains

¢

_____

¢

_____

¢

_____

¢

_____

¢

_____

¢

_____

¢

_____

¢

_____

Longer or Shorter?

Which do you think is longer,
your arm or your leg?
How much longer is it?

Which do think is longer,
your foot or your hand?
How much longer is it?

Which do you think is
longer, from the top of your
head to your belly button,
or from your belly button
to your toes? How much
longer is it?

Which do think is longer, from your
fingers to your elbow, or from your
toes to your knee?

COMPARING
MEASUREMENTS

61

🏠 **FAMILY NOTE:** Your child is using cubes to measure and then comparing measures
to find how subtraction can be used. Have your child tell what he or she found.

Name _____

 easuring Me

Find Out Which Is Longer	Estimate	Measure	Which is longer? How much longer is it?
		Arm _____ cubes long	
		Leg _____ cubes long	
		Hand _____ cubes long	
		Foot _____ cubes long	
		_____ cubes long	
		_____ cubes long	
		_____ cubes long	
		_____ cubes long	

Think About THIS

cubit

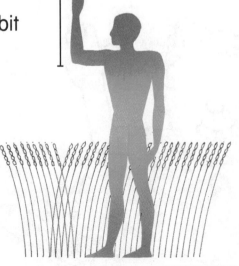

Did you know that long ago, Egyptians used a **cubit** to measure?

What can you find that measures 1 cubit long?

Growing Tall

Measurement	What I Measure	What My Buddy Measures	Difference
Height			
Foot			
Giant Step			
Arm Span			

🏠 **FAMILY NOTE:** Your child has been working to measure her or his height, length of foot, giant step, and arm span. Then your child measures the same attributes for a kindergarten buddy and compares. Have your child tell you about how she or he measured and used adding and subtracting to compare height.

Name _____

How much taller?

My Report

This is what I found when I measured.

This is what I think I may measure in Grade 4.

Think About THIS

Your hair grows about 15 cm each year. How long would your hair be if you did not cut it from the day you were born?

How many different ways?

JUICES
ORANGE
BANANA
CARROT
MIXED FRUIT

50¢ A CUP

🏠 **FAMILY NOTE:** For about the next week, your child will be investigating how to count collections of coins, how to show the same amount using different coins, and how to make purchases and make change. Ask your child to tell you about the coin sets shown at the juice bar, and how the sets are alike and different.

65

Name _____

Race to a Dollar

You need a pencil and a paper clip
to make the spinner.
Take turns with a friend.

- Spin the spinner.
- Take a coin like that.

Put your coins here.
Tell your friend when
you have a dollar.

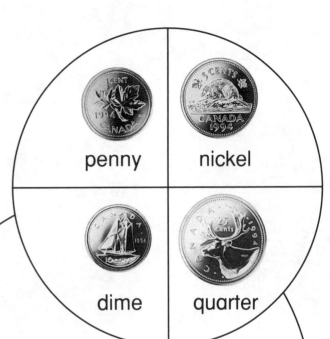

penny   nickel

dime   quarter

Think About THIS

Do you have coins in your pocket right now?
Don't look, feel the coins. Can you tell them apart?

Grooves around the edges of dimes and
quarters make them feel different
from pennies and nickels.
There are 118 grooves around a dime.
There are 119 grooves around a quarter.

🏠 **FAMILY NOTE:** Your child learned about using coins to make one dollar. Have your
child use coins to show you different ways to make one dollar.

My Store

Write the price for each item.
What coins do you need to buy each item?

FAMILY NOTE: Today your child worked with coins to find out how much he or she needs to buy these items. Have your child tell you how he or she decided on the price of each item and what coins would be used to buy them.

COMPARING
PRICES

67

Coins! Coins! Coins!

Look at the prices you wrote on page 67.
Draw the least number of coins you need
to buy each item.

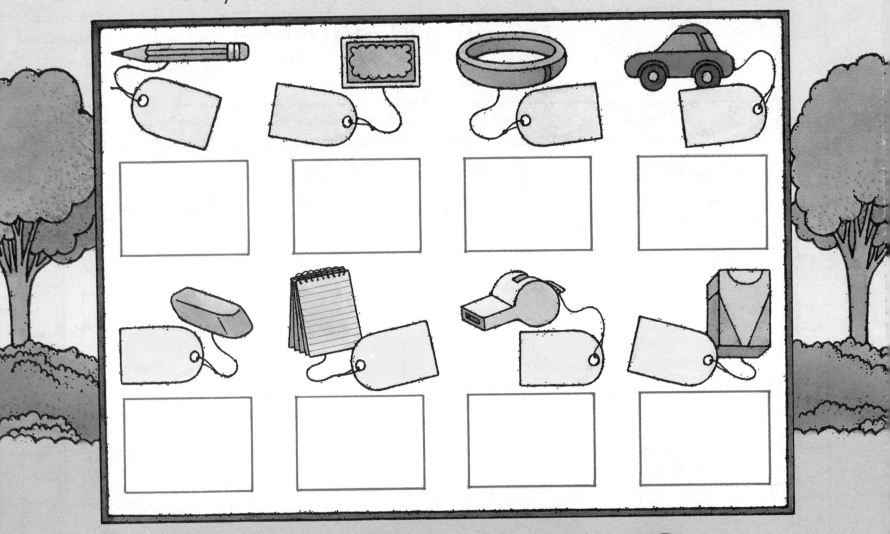

Think About THIS

Some of the first coins were shaped like beans.
Others were more like circles. They were made of
gold and silver mixed together.

Which is the biggest?

🏠 **FAMILY NOTE:** For about the next four weeks, your child
will be investigating ways to measure various attributes
of objects. The children will be developing language to help
them describe ways to measure and compare objects.
Discuss with your child the objects on this page and how
they compare.

69

Name _____

Class Olympics

GIANT STEP

Take a giant step.
Try to take
the longest step
that you can.

BLOWN AWAY

Place a Link-It on a table.
Try to blow the Link-It as far
as you can in one breath.

TARGET DROP

Drop a Link-It from
waist height onto a
target on the floor.
Try to get the Link-It
as close to the target
as possible.

🏠 **FAMILY NOTE:** Your child has been working to measure distances using materials
such as cubes. Your child is comparing his or her own measurements and trying for
personal bests. Have your child tell about the way he or she measured.

My Tries

Event	1st Try	2nd Try	3rd Try	Best Measurement
Blown Away				
Giant Step				
Target Drop				

Games

Think About THIS

A game like this one is played in Rwanda.
Rwanda is in Africa.
One person pushes the hoop.
The other person tries to toss a bean bag
through the hoop as it rolls by.
How far away from the hoop could you
be? Try it.

MEASURING
PERFORMANCES

71

Name _____

Dinosaur Data

11 m

Allosaurus
11 m long

15 m

Tyrannosaurus
15 m long

8 m

Stegosaurus
8 m long

9 m

Triceratops
9 m long

21 m

Apatosaurus
21 m long

FAMILY NOTE: Your child has been working to find what a given measurement looks like. Have your child tell you how the class worked to show the length of a dinosaur.

A Dinosaur in Our Room!

We chose this dinosaur to investigate.

_____

How long do you think it would reach?

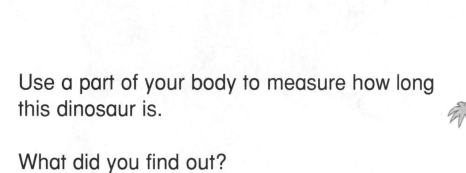

Use a part of your body to measure how long this dinosaur is.

What did you find out?

Think About THIS

The smallest dinosaur found so far was about the size of a chicken! It measured 75 cm long from its nose to the tip of its tail. It is called a Compsognathus. Are you taller or shorter than 75 cm?

MAKING
MODELS
OF
MEASUREMENTS

Dinosaur Dig

The earliest known dinosaur so far is called **eoraptor lunensis** (ee-oh RAP-tur loo-NEN-sis).

Eoraptor lunensis
About 100 cm long
About 69 cm tall
Back leg about 44 cm long

🏠 **FAMILY NOTE:** Your child has been working to compare measurements. Have your child tell how her or his measurements compare with that of an **eoraptor lunensis** dinosaur, and how he or she found out.

Dinosaur and Me

Think about how big **eoraptor lunensis** is.
Measure yourself.

Think.	Estimate.	Measure and compare.
Are you taller or shorter than this dinosaur?		_____ cm tall _____ cm tall
Are you wider than it is long?		_____ cm wide _____ cm wide
Is your leg longer or shorter than its leg?		_____ cm long _____ cm long

How do you compare with the dinosaur?

COMPARING
MEASUREMENT
USING
STANDARD
UNITS

Name _____

ake It Balance

A Hanger Balance

How to Make It

1. Use a hanger.
2. Make two chains with 5 paper clips on each chain.
3. Hook one chain on one side of the hanger. Hook the other chain on the other side. Use tape to keep the chains in place.
4. Hook the other end of each chain to a paper cup.
5. Hang the hanger.

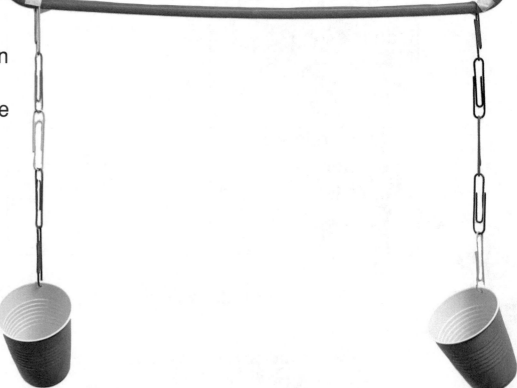

How do you know your balance works?

World Records

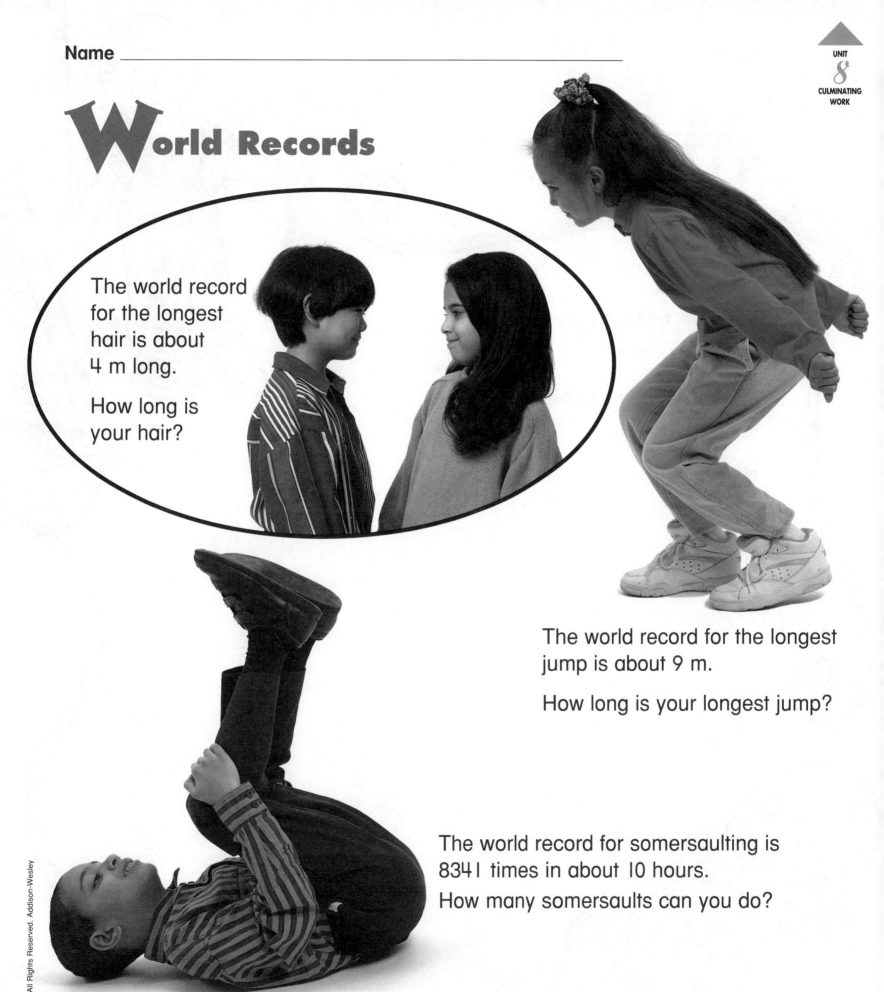

The world record for the longest hair is about 4 m long.

How long is your hair?

The world record for the longest jump is about 9 m.

How long is your longest jump?

The world record for somersaulting is 8341 times in about 10 hours.

How many somersaults can you do?

🏠 **FAMILY NOTE:** Your child has been working with the class to investigate class records for different attributes and events. He or she has been collecting measurement data for the attributes and events. Have your child tell you about how he or she measured and compared the measurements.

MEASURING
TO
COLLECT
DATA

77

Name _____

My Record

What I Am Measuring _____

What I Am Using to Measure _____

Data

What I Think

What makes an equal share?

🏠 **FAMILY NOTE:** For about the next two weeks, your child will be exploring fractions and probability concepts. Ask your child to tell you what he or she knows about equal shares, and what appears equal in the picture shown here.

Name _____

Cut the Cake

People are having parties!
How will you cut the cake to make fair shares?

🏠 FAMILY NOTE: Today your child learned about dividing a whole into equal shares.
Have your child tell about the cakes that he or she cut for each party.

Sharing Pizzas

You will be having a class pizza party.
Each of you will get one-third
of a pizza.

How many
pizzas should
you order?

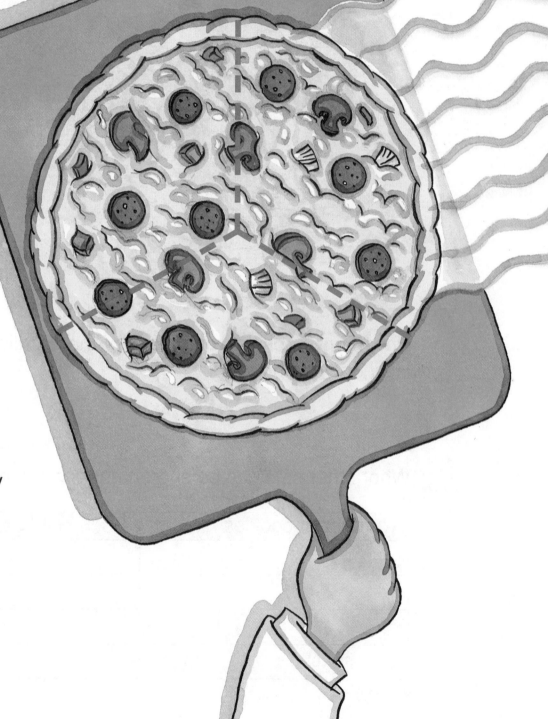

Write about how many
pizzas you need.
Tell how you know.

🏠 **FAMILY NOTE:** Your child has been finding how many wholes are
needed if each child in the class has an equal share. Ask your child
how many apples would be needed for your family if you each ate
one-half of an apple.

Name _____

Sharing Cookies

How many cookies would each of these children get?
How do you know?

What other sets can be shared equally among 3 children?

Number of cookies	Each child gets

🏠 FAMILY NOTE: Your child has been exploring which sets are easy to share among groups of children. Ask your child to tell you how he or she found sets that have no leftovers.

**ore on Sharing Cookies**

How many would each of these children get?

What other sets can be shared equally among 4 children?

Number of cookies	Each child gets

Spinning Spinners

Use a paper clip and pencil to make a spinner.
How many times will it land on each colour if you
spin 20 times?

Estimate.

Spin 20 times. Use tally marks to show on which
colour the spinner lands.

Write how many times it landed on each colour.

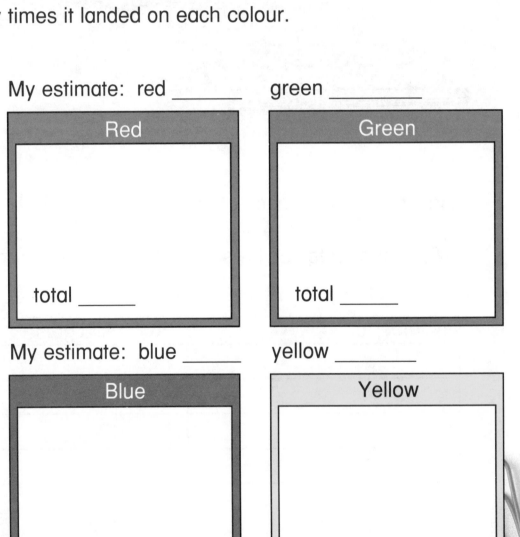

My estimate: red _____   green _____

Red	Green
total _____	total _____

My estimate: blue _____   yellow _____

Blue	Yellow
total _____	total _____

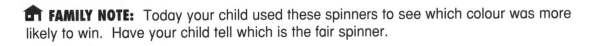

FAMILY NOTE: Today your child used these spinners to see which colour was more
likely to win. Have your child tell which is the fair spinner.

Finding Fair Spinners

Which of these spinners do you think is fair?
Tell why.

Spin the spinner.
Fill in a box above the colour you land on.
Stop when you fill in all the boxes for one colour.

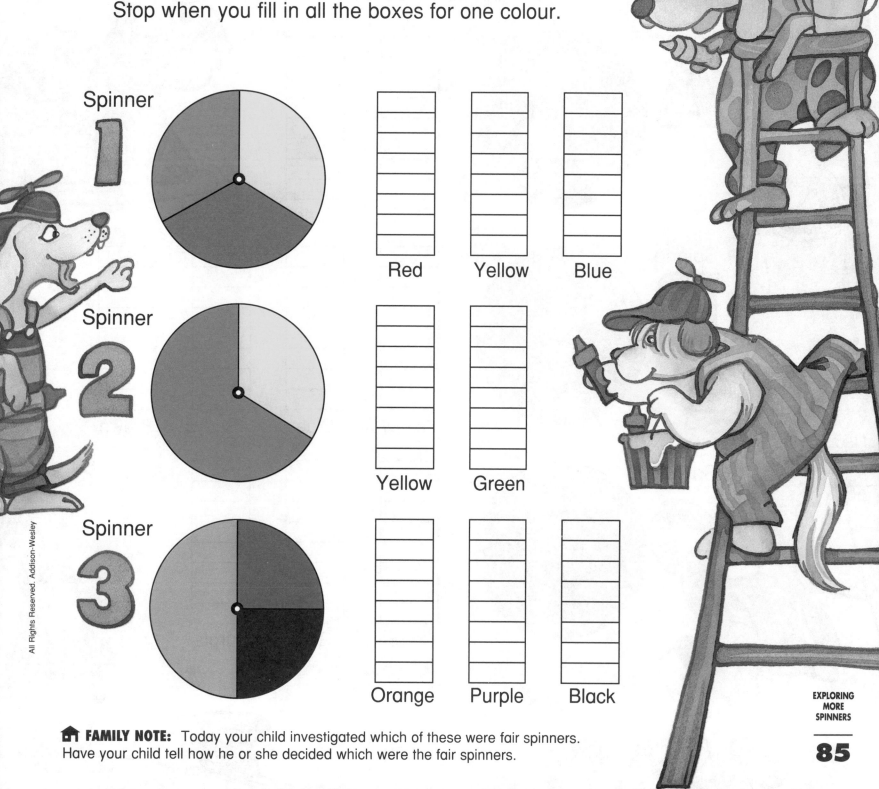

Spinner
1

Red   Yellow   Blue

Spinner
2

Yellow   Green

Spinner
3

Orange   Purple   Black

🏠 **FAMILY NOTE:** Today your child investigated which of these were fair spinners.
Have your child tell how he or she decided which were the fair spinners.

Name

Name _____

More Spinners

Which of these spinners do you think is fair? _____

Spin the spinner.
Fill in a box above the colour
you land on.
Stop when you fill in all the boxes for one colour.

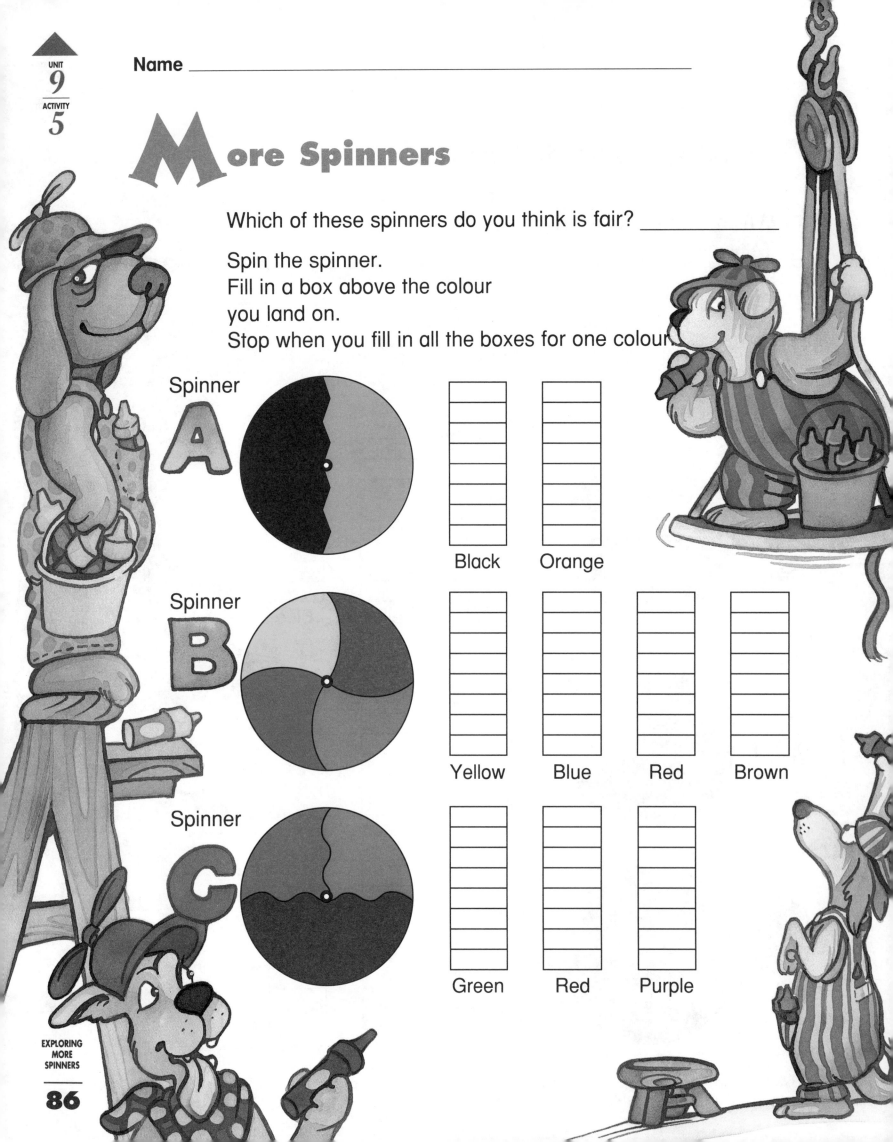

Spinner
A

Black   Orange

Spinner
B

Yellow   Blue   Red   Brown

Spinner
C

Green   Red   Purple

Toy Company Spinners

We still need spinners for these games.

Here are the spinners we need to make:
- A spinner that never spins red
- A fair spinner with two colours
- A spinner that is more likely to spin blue than red
- A spinner that spins green about half the time
- A spinner that spins yellow about one fourth of the time

Spinner I am making:

_____

Test Results:

Why my spinner works:

MAKING
SPINNERS

🏠 **FAMILY NOTE:** Today your child made one of the spinners listed. Have your child tell how he or she tested the spinner.

Name _____

Design Your Own Spinner

My spinner _____

_____

_____

_____

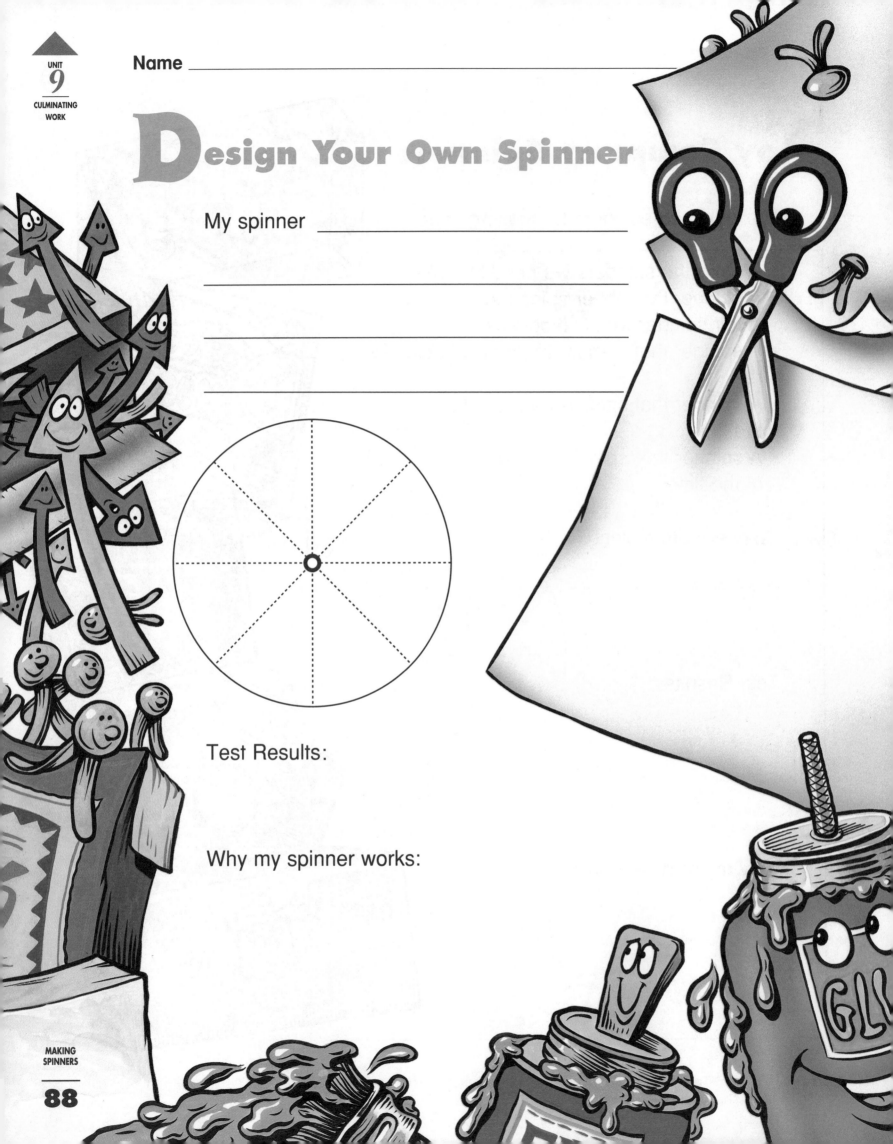

Test Results:

Why my spinner works:

How can we measure it?

🏠 **FAMILY NOTE:** For about the next two weeks, your child will continue to investigate measurement by having opportunities to measure the area and perimeter of shapes. Ask your child to tell you what kinds of measuring the children in the picture are involved in.

Name _____

Me and My Foot

I measured my foot.
This is what I found.

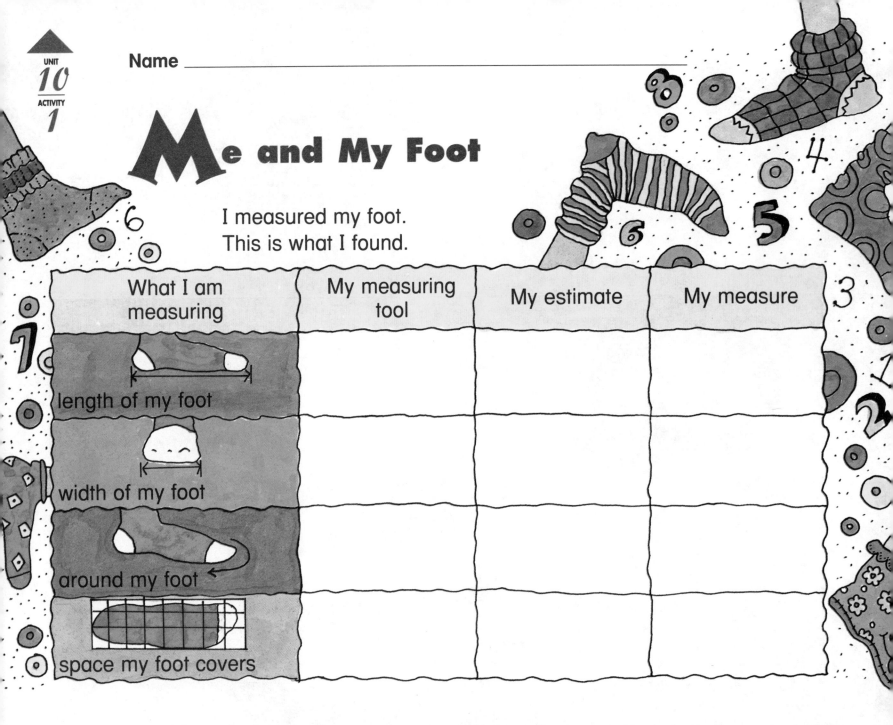

What I am measuring	My measuring tool	My estimate	My measure
length of my foot			
width of my foot			
around my foot			
space my foot covers			

Tell something you know about the size of your foot.

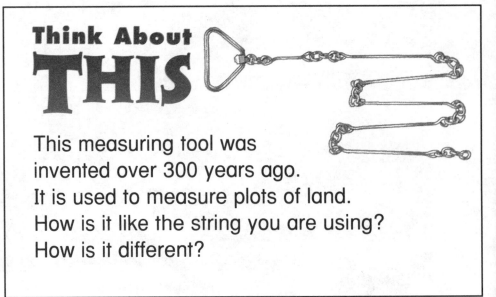

Think About THIS

This measuring tool was invented over 300 years ago.
It is used to measure plots of land.
How is it like the string you are using?
How is it different?

FAMILY NOTE: Your child has been using string and rulers to measure length, width, and the distance around a shape. Your child has also traced shapes on grid paper to measure area. Ask your child to tell you about these measurements.

Capture the Squares

This is what my shape looks like.	Estimate of inside squares	Number of squares captured

Which of your shapes captures the
most squares?

🏠 **FAMILY NOTE:** Your child has been forming shapes on grid paper, then counting the squares inside the shape to find the area. Ask your child to tell you about the figure he or she formed with the greatest area.

Tile This!

Use the smallest pattern block squares to tile each floor.
About how many squares will cover each shape?
Trace around to show how you placed the squares.
Record on Activity Master 42.

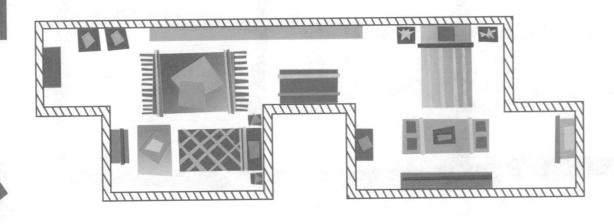

🏠 **FAMILY NOTE:** Your child has been working to measure the area of a space using square units. Ask your child to tell you how he or she measured the area of each floor.

Name _____

Tiles, Tiles, Tiles!

Large Tile

Medium Tile

Small Tile

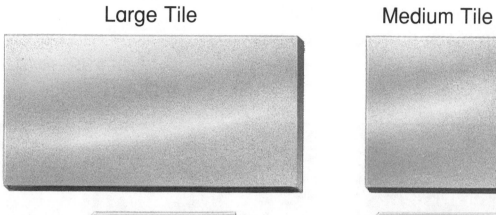

Large
Tiles
$10.00 each

Medium
Tiles
$5.00 each

Small
Tiles
$2.00 each

Tile the floor and ceiling using Activity Master 46.

Record your work here.

Floor

Which tile did you use?	Estimate how many tiles.	Estimate what you will spend.
	Count the tiles.	Find the cost.

Ceiling

Which tile did you use?	Estimate how many tiles.	Estimate what you will spend.
	Count the tiles.	Find the cost.

🏠 FAMILY NOTE: Your child has been using paper tiles to find the area of various surfaces. Ask your child to tell you how he or she "tiled" the floor and ceiling and found the cost of the tiles.

Name _____

Designing Tiles

People all over the world make tiles.
They come in many sizes and designs.

These tiles have Egyptian designs.

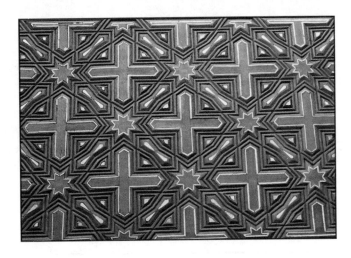

These tiles have Dutch designs.

Become a tile maker!
Draw your own designs on
these floor tiles.

Tiling Floors, Ceilings, Walls

This room has tiles on the floor, ceiling, and walls.
The tiles on each surface form a pattern.
Tell about the patterns you see.

Use the tiles on Activity Masters 43,
44, and 45 to tile your room.
Here are the costs.

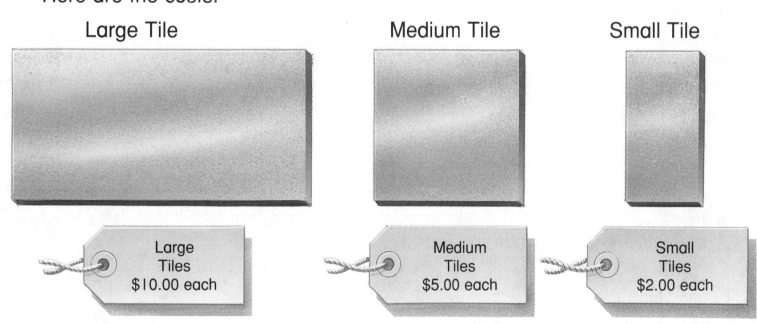

Large Tile	Medium Tile	Small Tile
Large Tiles $10.00 each	Medium Tiles $5.00 each	Small Tiles $2.00 each

🏠 **FAMILY NOTE:** Your child has been using tiles to cover a surface such as a wall or floor. He or she has formed a pattern with the tiles, then found the cost of the tiles using addition and subtraction. Ask your child to tell you about the room he or she planned, and how the cost of the tiles was found.

Name _____

Tiling Floors, Ceilings, Walls

Write about each surface you covered with tiles.

Surface	How many tiles?	Cost
	Total Number of Tiles	Total Cost

Write about what you found.

What equal groups can we make?

🏠 **FAMILY NOTE:** For about the next two weeks, your child will be investigating equal groups. This is the first step toward understanding the concepts of multiplication and division.

Name _____

Stories about Equal Groups

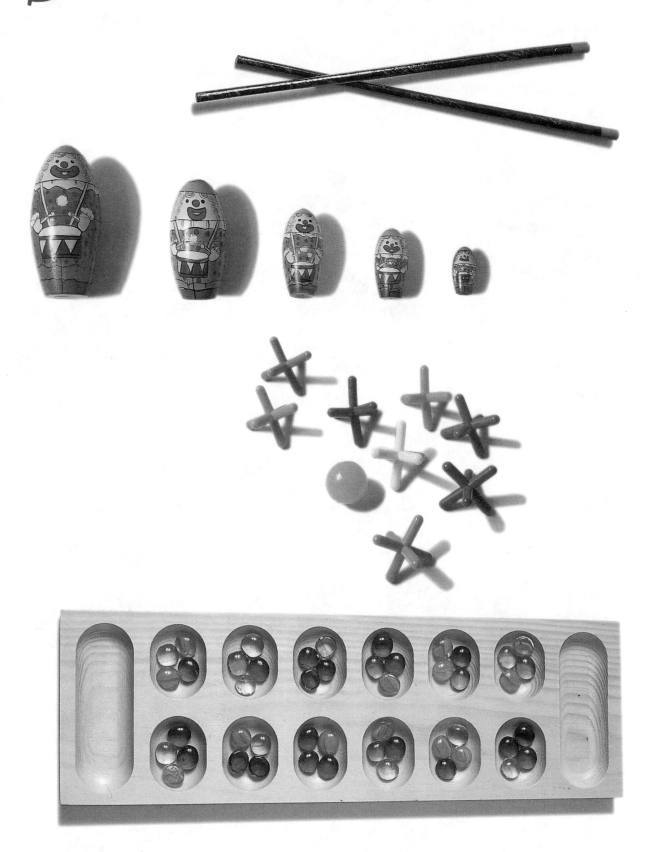

 FAMILY NOTE: We have been writing and solving problems based on equal groups of items. Here are some things from different cultures that come in equal groups. Ask your child to tell you a story problem he or she wrote based on these groups. You can create a story problem using these groups and your family.

People and Pets

How many legs are behind this fence?
Tell how many and show how you know.

⌂ FAMILY NOTE: Your child has been solving multi-step problems. Together, make up similar problems using family members and the imaginary pet your child drew on the next page.

COMBINING
EQUAL
GROUPS

Name _____

Here is my pet.

Make up a problem about the legs on your pet
and the legs on some people.

Draw your problem here.

Write your problem here.

The answer is _____

Sharing Treats

Here are the treats.

Make 3 equal bags. How many would you put
in each bag?

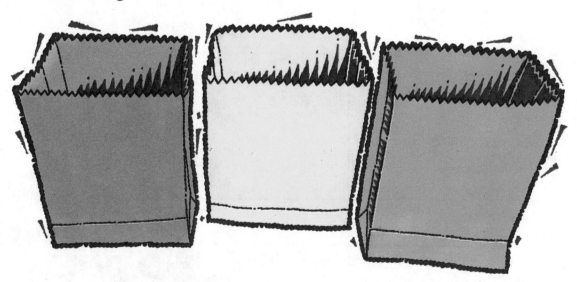

Were any treats left over?

How would you use the leftovers at a party?

SHARING
SETS

🏠 **FAMILY NOTE:** Today, your child learned how to share treats equally among
a number of children. Have your child find how many each of you in your family
would receive if these treats were shared in your home.

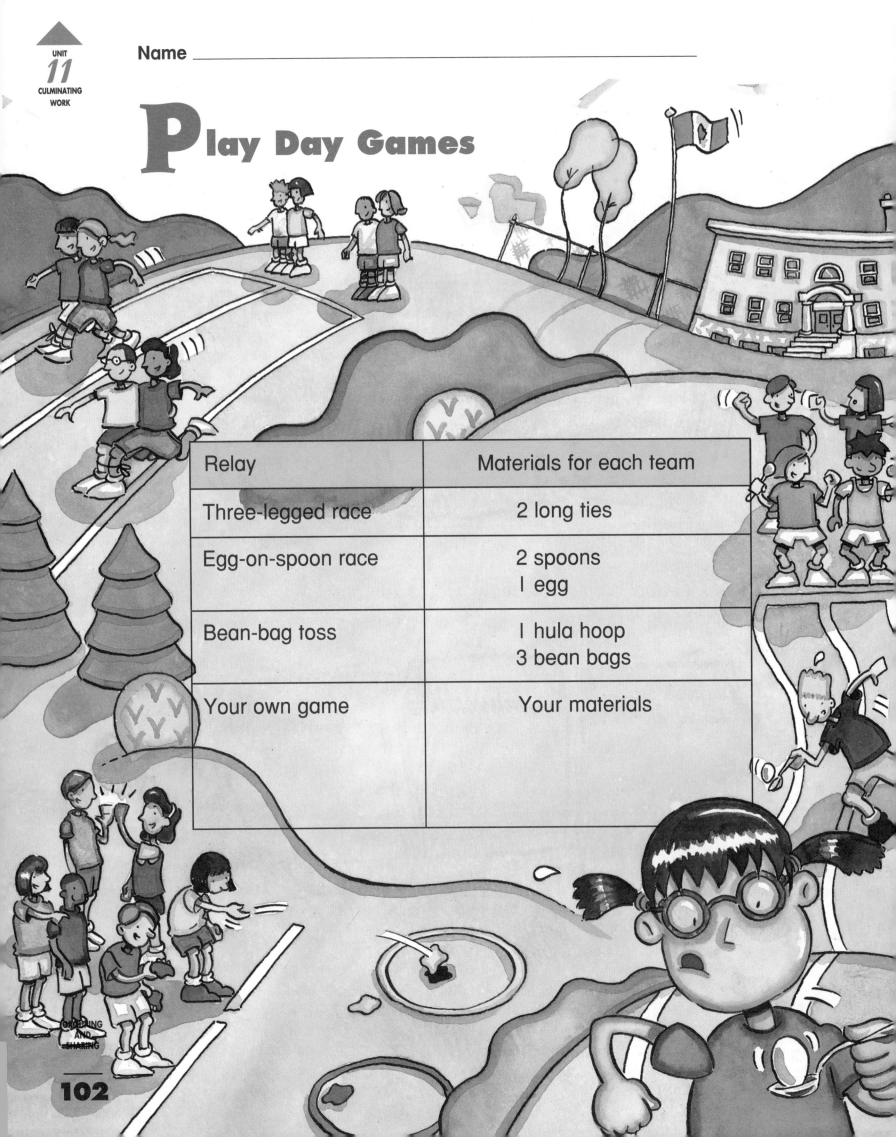

Name _____

Play Day Games

Relay	Materials for each team
Three-legged race	2 long ties
Egg-on-spoon race	2 spoons 1 egg
Bean-bag toss	1 hula hoop 3 bean bags
Your own game	Your materials

Acknowledgments

ILLUSTRATION

Cover Illustration: **Seymour Chwast**
Joe Bartos: 4, 5; **Bill Basso:** 24; **Alex Bloch:** 34, 62, 65, 68, 72, 73, 81, 90; **Joe Boddy:** 18, 19, 83, 84; **Bill Bossart:** 66; **Cindy Brody:** 55, 56; **Jane Caminos:** 90; **Adam Cohen:** 34; **Donna Corvi:** 54; **Lisa Cypher:** 53; **Luisa D'Augusta:** 2, 3, 48; **Robert Frank:** 74, 75; **Marvin Glass:** 52; **Obadinhah Heavner:** 69; **Steve Henry:** 8, 9, 28, 29; **Dan Hobbs:** 81, 102; **Dave Joly:** 57; **Brian Karas:** 13; **Elliot Krelof:** 30; **Linda La Galia:** 36-38; **Joe Lapinsky:** 93, 94; **Diana Magnuson:** 25; **Renée Mansfield:** 82, 83, 99; **Maria Pia Marella:** 78; **Claude Martinot:** 92; **Kimble Mead:** 61-64; **Mas Miyamoto:** 67, 68; **Gavin Owen:** 20; **Cindy Patrick:** 88; **Clarence Porter:** 21, 101; **Chris Reed:** 59, 60; **Doug Roy:** 14, 15; **Bob Shein:** 43, 49, 50; **Joel Snyder:** 45; **Linda Solovic:** 46; **Steve Sullivan:** 47; **Gary Torrisi:** 58; **Joe Veno:** 33, 80; **Randy Verougstraete:** 27, 31, 32; **Josie Yee:** 95, 96; **Jerry Zimmerman:** 16, 17, 85, 86.

PHOTOGRAPHY

Photo Management and Picture Research: **Omni-Photo Communications, Inc.**
Claire Aich: 1, 64, 79, 89; **Ian Crysler:** 23, 80, 97, 98;©**Culver Pictures:** 17; ©**W. Eastep/The Stock Market:** 42; **Everett Studios:** 53, 69; ©**Fotopic/Omni-Photo Communications:** 33; **Michael Groen:** 66; **Horizon:** 11,12; **Richard Hutchings:** 29, 30, 82; **Ken Karp:** 39, 40, 44, 65; **John Lei:** 26, 35, 51; ©**John Lei/Omni-Photo Communications:** 23, 66; ©**Loren McIntyre/Woodfin Camp & Associates:** 10; **Steven Ogilvy:** 7, 23, 32,71,76, 77, 88, 91, 92; ©**Shelly Rotner/Omni-Photo Communications:** 43; ©**Vince Streano/The Stock Market:** 42; ©**Sydney/Monkmeyer Press:**12.

CALCULATORS

T-I 108
T-I Math Explorer

MANIPULATIVES

Link-Its ™
Power Solids ™